Best Wishes
Judy Friday

This book is dedicated to Jack and Sue Tiffany

LOADSTAR
TIFFANY FARMS
LYME, CONN.

A Year at Tiffany Farms

by Judy Friday

Designed by Clare Cunningham Graphic Design
Essex, CT

Printed by Moore Wallace Andrews-Connecticut
Manchester, CT

Photo Credits:

page 14: photo courtesy of the Florence Griswold Museum; Lyme Historical Archives.
page 15: photo courtesy of the Florence Griswold Museum; Lyme Historical Archives.
pages 10-17: photographs courtesy of the Tiffany family
page 141: photograph by Hannah Shillawski

Introduction

I have always loved animals, and over the years I have spent a lot of time painting and sculpting the cows at Tiffany Farms. A few days before the end of 2002 the idea came to me to spend the next year photographing and writing about the local dairy farm in Lyme, Connecticut, in order to get a good record of it in all four seasons. The more I got involved with the project, the more I realized how much work is required to keep a dairy farm of its size going. Tiffany Farms is not considered a large farm, but it is not a small one either. Part of my motivation for recording the farm in detail is that I learned it is becoming more and more difficult for a small dairy farm to survive. Tiffany Farms used to be one of many such farms in Lyme, but over the years most of the farms have either ceased to exist as farms or have been sold. In 1940, there were 6,233 dairy farms in the state of Connecticut, and as recently as 1990 there were 356. Today there are 195.

The reasons for the disappearance of the farms are many. The costs of the land, equipment and feed keep going up drastically while the price paid to farmers for the milk is going down. Small dairy farmers are actually losing money every year now, as they are paid less per gallon of milk than it costs them to produce it. In order for small farms to continue they have to increase the number of cows they milk. Tiffany Farms milks a hundred head and covers 140 acres, but they can't get any larger because they don't have the land or buildings for it. As it is, they plant corn and other fodder crops in fields all over Lyme that do not belong to them in order to get enough silage for the cows they have.

I guess one could ask, "What would be the harm in small farms ceasing to exist in favor of larger farms?" Aside from the fact that the farmers would have to stop doing what they know and love to do, we, the people who live in an increasingly technological, fast-paced world, would lose the pleasure of seeing the open space and scenic views that farms provide. When a small farm goes under, the land is usually sold and more houses go up in its place. The value of these farms cannot be measured strictly in terms of the milk produced, but also from the peace of mind that comes from having them in our midst.

I hope to convey through photographs and writing not only how much thought and effort goes into the production of a gallon of milk, but also the spirit of the animals and the people who tend them. Some photographs in the book show the work being done on the farm and some are included simply for the fun of it, because there were many glimpses of humor and joy along the way. Obviously, one of my favorite photographs is the one of the cow poking her head out of a hole in the barn wall in order to get a glimpse of what's going on outside. This book is my glimpse at what's going on inside the barn wall as well.

History

According to the New England Historical and Genealogical Register, the Tiffany family in Lyme goes back to 1708, when Ephraim Tiffany moved to Lyme, Connecticut, from Block Island and, along with his brother Consider, bought a lot of land on Mount Archer and on Tantummaheag Road. Ephraim and Consider were the two youngest sons of Humphrey and Elizabeth Tiffany. Humphrey Tiffany came to Massachusetts from England around 1663 and died on July 15th, 1685, when he was struck by lightning while traveling by horse between Swansea and Boston. His wife then married Simon Ray, who had settled much of Block Island. Sometime in July of 1689 Simon Ray and Elizabeth were captured by pirates and held for ransom, and at that point Elizabeth decided Block Island was no place to raise a family; so she and Simon Ray and some of their children moved to Lyme.

The house and land where Tiffany Farms is today have been in the family since 1841, when Charles Ely Tiffany bought it from Captain Samuel Selden. I asked Jack Tiffany, the current owner, to tell me what relation Charles Ely was to him and he told me, "Well, we'll go backwards. There's myself (John J. Tiffany II), then Leon Tiffany, John J. Tiffany, the first of John J. Tiffanys, then there was David Tiffany, and then there was Charles Ely Tiffany, then there was Ebenezer Tiffany II, and then there was Ebenezer Tiffany, and then there was Ephraim Tiffany, and then there was Squire Humphrey Tiffany who came from England." Jack and Sue's children, John J. Tiffany III and Hannah, along with the children of Jack's deceased brother, Leon Jr., David, Jennifer and Andrea, make up the tenth generation of Tiffanys in Lyme. Out of all the generations of Tiffanys in Lyme I would say that Jack has probably seen the most change in the town. He was born January 6, 1932, and can remember quite vividly the 1930s and 1940s when he was growing up and can tell about a way of life that is by and large not seen any longer in this area.

Leon H. Tiffany, Miss Holman and a pair of Devon steer calves in 1912.

Above, four teams of oxen hauling wood. Right, cabin on MacIntosh Road.

Almost everyone had animals of one kind or another and most people had saddle horses. The Lyme landscape was wide open with few trees, as the town was full of small farms where the people raised chickens, pigs, cattle and horses. Oxen were used to work, and the Hamburg fair grounds was known to have some of the largest displays of oxen in New England.

Jack told me about the log cabin his father, Leon, who was a contractor, built on MacIntosh Road during the Depression to keep his crew busy during the slow time. Two mules named Doc and Sally were used to skid (haul) logs out of the woods, and the mason, Herm Bump, wrote a poem about the building of the cabin, which goes as follows:

Depression Days

Stewart had a hatchet,
Harry bought a saw,
Leon had lots of lumber.
But this lumber was in the raw.

Hal, he owned a broad-axe
And was pretty clever, too.
Herm, he done the scoring,
God, how Hal could hew.

Bill, he was the handy man;
Such men are very few.
He helped to chink between the logs
And build the chimney too.

Gramp, he was the teamster,
He and his bulls done fair.
Every time he'd get hung up
How he would cuss and swear.

Ed, he was the finisher;
You never heard him jaw.
He hung the windows, cased the thing,
And even laid the floors.

Now Doc and Sally done their share
They were a damn good team.
They pulled the logs and hauled the rails,
And never once were mean.

Things were rather slow that time;
There wasn't much money you see.
So we built the old Log Cabin,
And we called it the DIAMOND T.

We have workmen today who can build houses in the same manner as they did back then, but the logs are not being hauled out of the woods by a pair of mules or by teams of oxen as was the case then. Some of the roads weren't paved then and there was no electricity on MacIntosh Road as we know it until the 1950s, because hardly anyone lived on MacIntosh Road. The cabin's source of electricity was from big, glass Delco batteries, and they had some appliances that were run by gas. Jack, his brother Lee, and their mother, Gwen, would stay at the cabin in the summers. Jack's father would go to the farm on Route 156 during the day and come up to the cabin at night. His father would take one or two cows up to the cabin along with his five-gaited saddle horse, Rex. Jack told me, "I wasn't any bigger than Erica (Hannah and Dave's daughter). I was three or four years old and he would give me the job of leading the cow up to the cabin in the spring and back to the farm in the fall. Of course, he milked her by hand when he got up there and he had the animals who were all broke to lead then, because he showed 'em a lot. He went to Danbury to the big fair. They had a big fair in Danbury, and he used to show down there and Haddam Neck,

Chester and Hamburg. He went all over Connecticut showing his Devons, and of course they had these great big horns and they had the brass buttons on the ends of the horns." I commented that it must have been a lot of fun back then and he said, "Yes, yes it was. We used to have picnics and crank ice cream by hand, go on hayrides. Of course, you'd take your life in your hands today, you go on the roads with horses or wagon. I remember, this would be back in the '40s. Camp Claire over there, at the end of the summer, the highlight for the kids was to go on a hayride. That's when Dad had the team of horses and he would take the camp on a hayride about 5:30 or 6:00 at night or 9:00 in August or early September, and they'd go around Hog Pond or down to Ely's Ferry."

Above, Leon Tiffany and Rex, his five-gaited saddle horse. Left, Devon cows at hay bunk in the late 1920s.

While trying to locate a photograph for use in the history of the farm, Jack came across a letter written to his father by Edith Emerson which tells about the picnics they used to have in the late 1930s and how you could see Hamburg Cove from the hill above the farm:

July 4th 1963

My dear Leon:

Do you recall? Jack's first steer at the Hamburg Fair. A nice little boy and, I am sure, a nice little steer. I don't know the herdsman.

The locale is a bit hazy. It seems to be my field here. Altho' I don't know how come you were on it, because we hadn't built on it in those days. It is definitely "156" in the background and the fields are the "Peck's".

Anyway, it's a cute picture, and I thought it would amuse you on this 4th of July. Remember the picnics your mother always had on the hill over the way from your house? in the evenings? I remember how utterly beautiful the lower bay was, viewed from that point, in the sunset? How many of our happy memories are intertwined with your family!

Ever your friend,

Edith

The land was much more open then and allowed many long views of the surrounding landscape. The Lyme Impressionists who came to stay with Florence Griswold at what is now known as the Florence Griswold Museum would paint the open pastoral landscape and had close ties to the farmers and their land and animals.

Edward Volkert, shown on the opposite page, was known as the man who painted cows. He was one of the colony of artists who came to Old Lyme from New York in the summers to paint the pastoral landscape. What began as an occasional motif, cows in the pasture, became a sort of salvation for him when his wife left him and took their daughter with her, leaving him devastated. The country life in Old Lyme had a healing effect on him, and over time he found his life's work in painting the cows and oxen in the early dawn light on the beautiful open pastures around him. He said, "I have discovered many curious things about cows. They have a great individuality. When you go into a field, they come around you to investigate, and if you leave your easel a minute, they will lick all the paint off your canvas." Volkert particularly liked painting oxen and said, "Oxen are twice as good as cows at posing. The Old Lyme area of Connecticut is sparse, rocky farm country where ox teams were in use long after they were supplanted elsewhere."

Above left, Jack Tiffany and herdsman with his first steer. Below, Carleton Wiggins and companion. Opposite, Edward Volkert painting a pair of oxen.

Above, Leon Tiffany and Les Cone in front of the portable construction shed in the late 1920s. Left, Jack Tiffany with a pair of Guernseys in the early 1940s.

Above, Tiffany construction truck at the cabin in the 1930s. Left, Lee Tiffany with a pair of Devons in the mid-1940s.

Left, three yoke of oxen belonging to John Tiffany bringing in a jumper sled of wood in the late 1920s. Lower left, Doc and Sally, the pair of mules used to help build the cabin on MacIntosh Road. Lower right, John Tiffany's last pair of oxen bringing in corn, in front of the old chicken house and corn crib in the late 1930s.

Winter in New England has a spare, hushed beauty about it. After the brilliant oranges, reds and yellows of the autumn trees, it is a relief to see the bare, lean browns and grays. The trees in the distance at this time of year take on many colors depending on the level of moisture in the air and the temperature. They can range from mauve to blue but always with a hint of smoky gray. The farm is particularly beautiful at this time of year with all the open space covered in snow. It is a joy to see such a wide expanse of pure white with only small bits of color as accent. ■

Winter

Jack Tiffany sits down after his morning routine to eat his breakfast and read the newspaper from nine to ten every day. Sue, his wife, said this was the best time to catch him to ask about the workings of the farm. My first visit with Sue and Jack was on Sunday, January 5th, and I began to ask what would turn out to be over the course of the year a nearly endless list of questions which they very patiently answered one by one, with Jack occasionally snorting in surprise at my ignorance.

One of the first questions I asked was if I could photograph a calf being born, and Jack said, "Sure, if you happen to be around when it happens." I replied that I thought calves were born at a certain time of year, such as spring. He hesitated and then told me, "No, they're born year 'round, otherwise we wouldn't have a stable supply of milk year 'round." To him it was something so obvious that he couldn't help his momentary hesitation. Farmers are in control of when calves are born because the cows are artificially inseminated. There are no longer any bulls at Tiffany Farms.

One thing I've often wondered as I drive past the farm on a cold winter day is, "Do cows suffer from the cold?" On January 14th the temperature was 15 degrees and it wasn't expected to get much higher than 25 during the day. I asked Jack, "At what temperature do cows suffer from the cold?". He said, "Well, that's a hard question to answer. If it's a dry cold, they can stand the cold a lot better than humans can, but what the cows don't like is wet cold or a rainy, wet cold. You see in some of those pictures there, the cows are lying down in the snow, chewing their cud. You'd think that, you know, if you were lying down that you'd be uncomfortable in the snow, but they're not. Plus, those are a beef breed and they're more hardy than the dairy breed. Cows can go down to well below zero, but a caveat to that is they've got to be fairly well-fed because they burn a lot of energy to maintain body heat. A cow that's in poor condition, sick or rundown, is much more apt to succumb to cold than a healthy cow that's got some weight on. We feed them more in the winter than in the summer and we feed them a different kind of ration, a ration with more energy in it, such as corn. The dry cows can go down in the cedars or into a nearby barn to get out of the weather. They ought not to be out in a field that's completely open in extremely cold weather or windy conditions. The beef cows can go down to the lower barn and that's open, so they can go in and out at will and eat. Then they'll come up in the lot. They don't stay down there at the barn."

The heifers (young dairy cows) along with Cricket, their old horse that's at least 40 years old, are put into stalls in one of the barns for the night when Sue and her daughter, Hannah, feed them after rounding them up at 3:30 in the afternoon.

In order to keep the cows healthy with a good amount of weight on them through the winter, the farm buys a high grade of alfalfa from Idaho. On a 25-degree day in mid-January the truck from Idaho rolled into Tiffany's carrying 54 large bales of alfalfa weighing a total of a little over twenty-four tons, or about 49,000 pounds. The trip from Idaho to Lyme took three days, which included a minor breakdown in Iowa. The cost of shipping alone was $2,300, and the alfalfa itself cost about the same, meaning a six-month supply of alfalfa ended up costing almost $5,000. Jack went on to tell me that a load of cottonseed that came in earlier cost $4,000. I was surprised by these numbers and said, "I don't know how you keep doing it with all these prices." He said, "Well, we're living off depreciation. We can't continue for a whole lot longer doing that, too many buildings that need painting and roofs that need reshingling." He went on to tell me, "There's more milk than there's a market for, so that depresses the price." The price paid to farmers for milk right now is at an all-time low going back to 1971. It is a little over $13 for 100 pounds. The highest price paid during this period was $17. There used to be something called the Northeast Dairy Compact which was in existence from 1997 to 2001 and guaranteed a minimum price paid to farmers for their milk.

The Compact was done away with and a federal subsidy plan was instituted in its place. One plan in the works that could help farmers is "An Act Concerning the Fair Pricing of Milk" which would limit the mark-up on a gallon of milk that the supermarkets can charge retail customers. A gallon of milk sells for over three dollars and the farmer is only getting a little over one dollar per gallon. Prices in the stores for milk have never been higher, and the price paid to the farmers is at a 30-year low.

Unloading the alfalfa off the truck and getting it into the storage barn was not an easy task, as each bale weighs 900 pounds. Billy Firgelewski came to help, along with his brother Eddie. Billy would operate the front-end loader forklift, which belongs to Gary Reynolds of Reynolds' Garage and Marine. Eddie said he was there generally to supervise and cart off the dead bodies after the unloading was over. The bales were apt to tip over; so a tractor had to prop them up on one side of the truck while the forklift picked them up from the other side. Dave Shillawski, Hannah's husband, and Eddie oversaw this part of the operation, and Kenny Stewart worked another forklift. Jack had the bales stacked in the barn. There was a certain sense of excitement in watching the unloading because these bales of alfalfa were so big that one of them really could kill someone if it happened to fall off the truck. It was also amazing that the alfalfa came all the way from Idaho. The trucker drove through the New York city area on his way here, which he said he'd never do again, and the truck loaded with alfalfa must have been quite a sight on the roads there.

Winter on the farm can mean more work if it snows on a consistent basis. In the winter of 1995 it snowed from Thanksgiving to March on a practically weekly basis. This winter has been just as bad, actually worse, because it's been so cold for weeks on end with no breaks in the temperature. On February 17th, it snowed 17 inches and the weather forecasters called it winter storm Daniel. A week later we got another foot of snow. Keeping up with the shoveling and plowing of all the paths and barnyard areas is the difficult part of the winter season. Ice forms on all the flat surfaces and it has to be chipped away on the warmer days. Cows can slip very easily on slick surfaces. If their udders get knocked around or kicked by another cow, there will be blood in the milk and it has to be discarded.

Dreamboat is the oldest cow at the farm, but she is still giving a good quantity of milk at 13 years of age. As she was walking by one day, Jack commented "She moves slowly, she always has." I said, "Maybe that's why she's lived so long," but he quickly corrected me, saying, "No, she's lived so long because she keeps having calves and giving milk." Not too many cows are kept around for sentimental reasons, although Dreamboat will be kept on the farm as a pet until she dies of old age. The Tiffanys also have a blind cow, named Popsicle, who shares a field with the old horse, Cricket. She was born blind and they kept her as a pet. She has calves and gives milk like the other cows, but she needs to be led to her milkings, the field and her stall for the night. When Bridget, Kenny's daughter, goes to get her out of a stall and lead her to the field in the morning she sings to her. She sings because she said, "Popsicle isn't sure who I am until she hears me sing." I asked her what Popsicle's favorite song is and she said "Stairway to Heaven."

On Wednesday, February 12th, I went to help Sue feed the young calves. She told me a delivery of sawdust was coming in that morning. Just as the truck full of sawdust pulled in a big snow squall kicked up. The wind was making the snow come down sideways, so when the sawdust was pouring out of the back of the truck you couldn't tell if it was sawdust or snow that was swirling around.

The farm goes through tons of sawdust. The delivery that day would last a month and cost $1,000. Terry Joseph, the driver of the truck, told me this was good sawdust. I didn't know that some sawdust was better than others. He told me that for the farm's purposes, a coarse sawdust made from kiln-dried pine, is better than a fine sawdust made from oak or maple, because the fine sawdust creates dust which can cause respiratory problems in the cows. More sawdust is used in the winter than in the summer because the cows are inside the barns more in the winter. The barn floor and the stalls are cleaned each morning and fresh sawdust is put down as bedding after each cleaning.

Needless to say, once the sawdust was unloaded and stored away, the sun came out and the snow stopped. The wind didn't stop, though, and more than once I've heard Jack say he doesn't mind the cold so long as the wind doesn't blow. I now know what he means. A thirty-five degree day is going to feel like spring after the winter we've had.

When you drive down Route 156 past the farm in the morning, it looks, from the road, like not much has to be done to keep the cows you see lying in the fields fed and watered. You'd have to go into the barns to see that there is a routine that is carried out from 5 a.m. until noon every day that is very much orchestrated as to what gets done when and in what order. One job can't be done until the necessary job before it gets done. Looking at the barns from the road is like looking at a beehive from a distance. From the outside it always looks the same, but on the inside everyone has his and her job and all the jobs are related.

One of the jobs is spreading sawdust in the loafing barn. The loafing barn is a large, open barn where the cows lie down and chew their cud after they've been milked and fed. Kenny Stewart, who has worked at the farm for the past 16 years, pushes the bin of sawdust, which hangs from and moves around the barn on tracks set in the ceiling in a horseshoe pattern, and shovels the sawdust out as he goes. There are a lot of systems in place at the farm which make the jobs much more efficient to carry out. This is one of them.

NAPA
NAPA

Repairs on a farm are required constantly, but the season to catch up on them is winter. The field work is finished until spring and there is finally a little spare time. In early January before it turned bitter cold for months on end, John, Jack and Sue's son, who does most of the carpentry work on the farm, built two new gates, which were put in along with new fence posts. Jack, John and Kenny then replaced the old barbed wire with new wire. The wiring is done very quickly and efficiently. A large spool of barbed wire is spun out and tightened at the end of a row of fence posts before it is nailed onto each post.

I didn't realize how important it is to have someone around who can fix things on a regular basis. John, in addition to his other chores on the farm, does most of the repair work. For fun, I asked him for a list of the repairs and new construction that he did in the month of January.

They were:

- fixed three stanchions
- made three new stanchions
- built two new gates
- fixed bathroom
- tiled milk parlor
- built new feed bunk
- repaired faucet in calf barn
- put paneling on wall outside of bathroom
- helped with barbed wire on fence around fields
- put in support post for back wall at lower barn
- repaired light at lower barn
- repaired porch panels in front of milk parlor
- repaired post supporting big bale of hay above cow hospital
- replaced turn signal on dump truck
- serviced portable heater
- fixed antifreeze leak in feed truck

This is what he could remember doing, as I asked for the list in late January. I'm sure there were more, as well as the work Jack and Kenny did.

10-15-01
Peace

Yesterday, March 21, was the first day of spring on the calendar. It had rained all day the day before and the cold air was still coming out of the ground which, when it met with the warmer air temperature, produced what they call in Ireland a “soft” day. It was foggy. The grass was just beginning to green up but the trees had not yet turned that wonderful shade of red that happens when the buds are about to pop. Starting around March 15th this year, the temperature warmed up considerably and it felt quite spring-like, although it would only last a week before reverting to cold and overcast. In New England, there is usually a long wait between the first day of spring and the first spring day. This year was no exception. On April 7th, we got another blast of winter with temperatures in the low 30s and more snow. ■

Spring

The temperature today, March 24th, was up around 55 degrees; so Hannah and Dave decided to put the young calves out in the penned area where they can run around. In good weather the calves are let out in the penned area for the day and put back in their warm hutches for the night. It had been too cold for them to be out of their hutches until today, so it was their first time out, and it was great to see them run around and kick up their heels. Some of their kicks and jumps were pretty comical, as though they were literally jumping for joy.

One of the young calves was given the name Present, as she was born on Hannah's birthday. Calves are up and walking hours after their birth and some look pretty big just days after they're born.Present was six days old on March 24th and still a little unsteady on her feet at first when Hannah went to lead her out of her hutch to the penned area.

Once or twice a year the cows need to have their hooves ground down. It is not an easy task. First, the cow is strapped up and lifted hydraulically onto a table, which puts the animal on its side. From there Dave takes a knife and cleans out all the dirt and manure so he can see what he's working with. Then he uses the trimming wheel, which is a high-speed trimmer with five small blades on it and takes small amounts of the hoof off at a time. He told me, "I basically shape the foot with the trimming wheel, take material out from the bottom of the foot, make it smooth and slope it so that the heel is high and the toes are lower. Then I shape the outside of the foot, get a nice curve, and after that I take the double-sided knife and clean between the toes and shape that a little bit and clean out any erosions or anything of that nature. After that, it's just about done. I take a rasp and clean up any sharp edges on the side of the foot, trim it, and back down she goes."

When watching this process you realize how immense cows are. A mature cow weighs about 1500 pounds. If they happen to twitch or move a leg suddenly, it's quite a jolt. I've observed a lot of chores being done on the farm, and this is one I would especially not want to do myself. Dave is able to trim the hooves of four cows in about two-and-a-half hours, each cow taking a half hour, and another half hour to round them up and strap them in. He tries to do the hooves of four cows at least once a month, but if he's behind from, say, the rough winter we just had, he'll do it a couple of times a month to catch up. He can tell which cows need to have their nails trimmed because he gets a good look at them when they come in to be milked and their hooves are nearer to eye level.

As is the case with a lot of things we embark on, if we knew what was involved before we started a project, we'd probably never begin. The painting of the gray barn near the house, which is called the horse barn as it was used to house the horses before Jack was born, was begun in early April. The task of scraping and sanding the old paint off before two and sometimes three coats of oil primer and the top coat were applied was begun with great enthusiasm by Andrea Tiffany, Jack and Sue's niece, myself and Jack. As the work progressed, windows were replaced, the barn's interior structure was bolstered in places and some outside boards were replaced. The painting and repair work of the horse barn took the month of April and part of May to complete, with three to four people working on it at different times, weather permitting. It was a relief when the final coat was put on, but the painting of one wall of one barn is a drop in a bucket when so many structures need painting. Maintenance of the barns and houses, which number twelve in all, requires constant repair, painting and reshingling. John does most of the repair work, and Jack does most of the painting.

Brent Kirby came to the farm one day to trim Cricket's hooves. Cricket is Hannah's horse, which she has had since she was 16. It was a warm April day and the farm was buzzing with activity. Brent has been a farrier, or horse shoer, for most of his life and has been taking care of Cricket's hooves for most of Cricket's life, which is over forty years long. While Brent was starting to trim, Jack and Brent were talking about a man from Middletown. I was only catching bits and pieces of the conversation; so Jack later told me, "When I was a kid, we had a team of big draft horses and we used to get this fellow down from Middletown who shod them and he had a forge on his truck and he would make the shoes right there, what they call 'hot shoeing', and that was a big thing in the 40s and 50s. Most farms had teams of horses as well as a small tractor and so a lot of the work was done with horses, which meant there was a lot of work around shoeing horses. Today, nobody has big draft horses except for hayrides or parades or the pulling contests; so, you know, whereas there were hundreds of teams of horses in Connecticut in the 40s, there may be a hundred teams today, if that."

1-26-03
Dalila
Ariel

By the end of April the frost was sufficiently out of the ground to begin spreading manure in the fields. The manure that had been accumulating over the winter was loaded onto trucks and hauled to and dumped in various fields on the Tiffany property and in other fields in Lyme, East Lyme and Salem that the farm also works.

Spreading manure is a two- or three-man job, depending on how many people are available to work. One person, in this case John, works the pay loader and fills one spreader with manure while the second person, Jack, drives a tractor with a full load of manure in the spreader until it is empty. Jack then switches from the tractor with the now empty spreader to the second tractor with the full spreader of manure that John has been filling while Jack is spreading. If a third person is available, Jack would operate one tractor and spreader and the third person would work the second tractor and spreader. The job is much easier with three people than two, as with only two it's a lot of up and down from one tractor to another in the course of working one field.

NEW HOLLAND

Before the spreading of manure can be done, rocks are picked out of the fields. The soil in Lyme is very rocky and new rocks come to the surface every year. That's where all the stone walls in Lyme came from. When farmers were clearing their fields for planting, they took the rocks and built walls around their pastures. Billy and Eddie Firgelewski came to help Jack, John and Kenny dig out and move large rocks that had risen to the surface over the last two or three years. They moved some dirt around at the same time, leveling the field for the repeated rounds of harrowing that would be done prior to and after the spreading of fertilizer by truck.

According to the book *Lyme Yesterdays*, by James E. Harding, written about life in Lyme in the mid-1600s, "The town of Lyme embraces 42 square miles 'more or less,' as the old land deeds say. For the most part the terrain is a series of steep, rocky ridges tending east and west. Solid outcrops of rocks and glacial deposits of boulders are so closely interspersed that one may walk for miles without setting foot on soil. In all the 42 square miles there are probably not more than 250 acres of prime arable land."

ATHENS
76

Once the rocks are picked and the manure is spread, the fields are harrowed. Harrowing turns the soil over, working the manure and rye that is planted in the fall to hold the fields together over the winter, into it. The preparation for a field of alfalfa is different from that for a field of corn. Jack explained the difference to me, saying, "Corn needs a lot of nitrogen and manure is rich in nitrogen, plus manure has weed seeds in it, and it's easier to control the weeds in corn than it is alfalfa; so we put the manure on the corn ground. Alfalfa doesn't need nitrogen; it needs potash. So we don't put manure on the field of alfalfa." It varies from year to year how much corn and how much alfalfa is planted. To participate in a government program the farm must submit a plan of conservation to the federal government. The plan includes a map of the farm showing the location of every field, the size of the fields, the soil type of each field and the degree of slope. Jack told me, "The government has a lot more say and control over our agriculture than it did thirty or forty years ago. They're most interested in anything that's considered highly erodable, what they call HEL land. We do a soil-test on a third of the farm's fields every year so that the entire farm gets soil-tested once every three to five years. Once the samples are sent in, they come back and tell you how much lime and fertilizer to apply. In the winter or early spring you kind of have to make up your mind on a crop plan, and you decide what fields are going to be rotated, according to your plan of conservation, according to your needs and according to your financial situation. You have to put everything into the mix."

This year the field by the firehouse, which was planted in corn for the last three years, is being prepared for the planting of alfalfa. Since this field has some slope to it, a sod crop such as alfalfa had to be planted there in place of corn so that the land won't erode. The farm is planting 80 to 100 acres of corn and 10 to 11 acres of alfalfa this year. Preparing a field for alfalfa is time-consuming and ends up being very expensive, but as Jack said, "Hopefully you can keep a stand of alfalfa for six to eight years before it runs out and becomes more weeds and poor grasses; but occasionally it will winter-kill or for whatever reason it gets diseased or an infestation of bugs and you have to plant it sooner. Theoretically, you're supposed to rotate your crops, have a sod crop for five or six years and then a row crop for a couple, then back to sod." Sue Tiffany, who pays the bills at the farm, told me, "We spent close to $20,000 on fertilizer, seed and herbicide spray for the crops last year. That's a lot. Every time the truck comes in the yard I cringe."

From April until October Jack, John and Kenny will be working in the fields beginning with picking rocks, spreading manure, and planting and cutting various crops for the feeding of the cows. The amount of work and money that goes into the fields is something that I find mind-boggling, but I'm so grateful that they like what they're doing enough to do it year after year. Some farmers are fussier than others when it comes to preparing the fields for planting and Jack clearly cares about aesthetics. The piece of ground by the firehouse has a gentle slope to it, rolling down to the adjacent flat field that will be planted in corn. The bare fields already look beautiful, but will be quite something when the crops come up.

Lucinda, the only Guernsey cow at the farm, gave birth to a beautiful calf named Lucky in late April. Hannah bought Lucinda for her father, Jack, a few years ago because when Jack was young he had a small herd of registered, purebred Guernseys. When Jack's father, Leon, started dairying in 1941, he had a mix of cows, with some Ayrshires, some colored cows of various grades and some Guernseys which were not registered purebred. In the early 1940s, when Jack was nine or ten years old, his father bought him Flossie, his first registered, purebred Guernsey, and over the years Jack built up a small herd of Guernseys. Guernseys produce a milk high in butterfat content, and by the early 1960s the market for milk had changed to one where the demand was for low-fat milk, which Holsteins produce. Jack's father went to Canada and bought six or eight Holsteins and gradually switched over to milking all Holsteins, and the Guernseys were eventually sold at auction. It was Jack's job to take his Guernseys to auction in Pennsylvania, and he said, "It was a sad day for me." It ended up being a sad day for Hannah recently too, as Lucinda had to be put down due to a compaction of sand in her stomach. The vet operated to try to remove it but Lucinda didn't recover. It's nice to see Lucky at the farm with her big, beautiful eyes and ginger, red and white markings.

It's the second week in June, not yet summer, which begins the 21st of June, and the weather has been foul for most of the spring. Although the flowering trees and shrubs were exceptionally beautiful this year, we've had rain, rain, and more rain. We get maybe one sunny day a week, but not more than two in a row. The farm now really needs three straight days of sun in order to make hay. There's that old saying, "Make hay while the sun shines." So far this spring they haven't been able to make hay because there hasn't been enough sun. They're out of hay and if we don't get sun soon, the grass will go by and won't be good for hay until the next mowing. Last year the weather was so dry that the hay production was down due to lack of rain. Mother Nature has more than made up for it this year.

Jack was able to mow some grass this week at the Nature Conservancy field near the farm on a rare sunny day. The grass will be chopped and put in the silo for silage instead of baled for hay. I asked if this couldn't take the place of hay and Sue told me that the cows like routine as much as we do. They're used to eating hay, which fills their stomachs. The grass that is used for silage is mowed and chopped in the course of one day versus the grass that is mowed for hay, which needs to sit and dry out for three days before it is baled. I was told that if the grass that is cut for hay doesn't dry out and is put in the barns it could spontaneously combust.

The chopper and the truck that catches the chopped grass must drive side by side so that the grass goes into the truck. It's then taken back to the farm and put into the silo. The next day the rain came back, but according to the weather forecast the rain will stop in a couple of days and we'll have sun for three days in a row; so then they can get to the season's first haying.

In the meantime, Kenny is working on the hay-baling machine, which was bought in the 1970s. There are a lot of machines that need to be tuned-up every spring for the various aspects of the field work. These include the manure spreaders, the harrows, the seed planting machines, the hay rake and the hay baler, not to mention all the tractors that pull these various machines and the big trucks that haul manure and carry the equipment to the outlying fields spread around Lyme, East Lyme and Salem. If one ever needed to learn that we're not in control of everything, farming would be a good way to realize it. They do what they can in the fields on the good weather days and repair or tune up equipment on the bad days.

We finally got some heat on June 23rd. Usually, by this date, there have been many days of temperatures in the high 70s and 80s, but this year the summer heat arrived right around the same time as the summer solstice, which was June 21st. After slogging through one of the rainiest Junes on record with 5.07 inches of precipitation as of June 23rd, the sun finally broke through. The beef cattle like to sit in the shade up by the side of the road on these hot summer afternoons. They sit with their eyes closed, chewing their cud, and by and large stay still long enough to set up an easel and do a sketch. Observing the cows with their quiet demeanor never fails to have a calming effect, and I'm thankful to be able to spend an afternoon this way. ■

Summer

On June 25th, the season's first hay was baled, although most of it would be used as mulch hay instead of feed hay as it got too wet after it was mowed.

There are many steps to haying. First, the grass is mowed with a machine called a mower-conditioner. The grass gets mowed and then goes through two rollers that turn at very high speeds, and the rollers crimp the stems every one-and-a-half to two inches. Jack explained that, "The crimping makes the hay dry a lot faster, so conditioned hay will dry in two days, whereas if you just mowed it, it would take a lot longer to dry. The mower-conditioner is one of the dramatic changes in the making of hay in the last 50 years. The first big change was from horse power or ox power to tractor power. The haybine (modern mower) mows nine feet at a time, almost twice the width of the cutter bar that the horse team pulled. Right after the war they developed the crimpers which were originally two steel rolls, then they came up with rubber rolls, and then they came up with the baler. So there's been a lot of change in 50 years."

After the grass is mowed, it's raked into windrows with the hay rake, then it's tedded, which means it's shaken and spread out for drying, before it's turned over again with the hay rake. The next day it's tedded and raked again before it's finally baled and loaded onto the trucks.

HAYLINER 283
SPERRY RAND
4020

TIFFANY FARMS
LYME, CONN.

The baling machine is thirty years old and is a pretty incredible machine. The hay gets sucked into the front end and is pushed out the other end in bale form, with strings around it that are already tied. Haying requires a lot of manpower and machinery from start to finish. In addition to the people who work the farm on a regular basis, Erv Dill drives the trucks that get loaded with the hay and has been helping with the field work for many years. He also gives the farm sawdust shavings from the bird houses he makes. Two high-school-age boys usually help with the loading of the trucks. The two fields that were worked on June 23rd to June 25th are about 10 acres and yielded over 500 bales of hay. The farm puts up 15,000 bales of hay over the course of the spring, summer and autumn to see them through the year. Although most of the hay from these two fields would be used as mulch, some feed hay was baled also. It is a relief to have the beginning of a supply of feed hay at the farm.

While the field work is done seasonally in order to feed the cows, the breeding of the cows continues on a year 'round basis in order to milk roughly 100 head consistently. Breeding is carried out by artificial insemination at the rate of 10 to 15 cows a month. Everyone at the farm has a unique role in keeping the place going. Dave Shillawski, Hannah's husband, and Hannah oversee the breeding; Dave performs the artificial insemination, but they are both very much interested in improving the herd. One of the ways to do this is by having the cows classified in order to identify the cows with the desired traits and then "flushing" them in order to implant their embryos into cows that are thought to be of a lesser quality.

Devlin Hurley, who is a professional classifier (rater), came to the farm to rate the cows and explained to me what the rating system is about as he rated Millie. He said, "This is Millie. She's a little stale, but a pretty good cow, by the look of her." I asked what stale means and he said, "Stale means it's been a long time since she had her last calf. When a cow first has her heifer or calf we say she's a fresh cow." Dev went on to say, "Millie's three years and four months old; she calved last January. She's had one calf. The first thing we rate is stature. This cow is a fairly tall cow, probably just shy of 59 inches, so she's about a 42 in stature. [This is on a scale of 1-50]. Strength, we're looking at the fore-rib and the width between their chest floor. You know, she's a little above average in strength kind of cow, I would say. So we're going to make her 33 in strength." He explained many more categories of rating, such as body depth, rib angling, slope of the pins, rear view, foot angle, fore udder attachment, rear udder height and width, center support, teat placement of the front and rear teats, udder tilt and teat length. I could see from one of the photographs that, as Dev was rating the next cow, Millie seemed to be checking out his rear view. Those categories were called "linear descriptive" and they're rated on a scale of 1-50. The next rating is done in five major categories on a scale of 1-100, but Dev said they really only use from 50-100. Dave jokingly said, "If a cow rated below 50 it would basically have just four legs, a head and say moo." Millie rated an 85 in frame structure, 86 in body capacity, plus 80 in dairy character, 78 in feet and legs and 83 in the udder. So she came up with an overall rating of 82 out of 100, which is pretty good. A rating of 97 is as high as they go. Dev rates cows pretty much every day, although he travels a lot, as his territory covers Maine to North Carolina, with occasional trips to California, Washington, Michigan, Kentucky, Ohio and West Virginia. He rates only Holsteins.

Rating also enables the farm to sell the cows more easily, as it determines their value relatively. One of the cows, named Sleet, was given an excellent rating of 90 when Dev was there, so Dave and Hannah decided to "flush" her. Flushing is done by a veterinarian, in this case Kevin Lindell, who comes to the farm on a monthly basis to check pregnancies and perform flushes. Flushing is quite an involved procedure, although painless for the cow. The point of the procedure is to produce multiple eggs in the highly-rated cow through super ovulation. The cow is inseminated, which fertilizes the eggs, and the resulting embryos are "flushed" or removed from the cow for transplantation into cows of lesser quality, thus extending the particularly prized cow's genetic traits far beyond her single annual offspring.

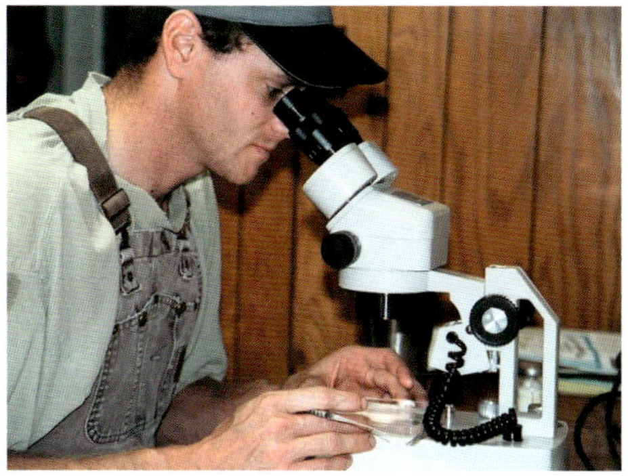

In June Kevin and his assistant Jennifer Lucy flushed Sleet and explained the procedure to me as they went along. The first thing they do is give the cow an epidural which numbs her hind quarters. He then puts a long glove on his left hand and arm, inserting his hand and arm well up into the rectum, removing manure. He said, "The uterus is right below my hand, we manipulate it and guide this catheter up in there through her cervix into her right horn. All of her ovulations are on her right horn ovary. So, I'm trying to flush just the right horn. We're flooding her uterus right now with this flush media and then we recover it through the same catheter." The flush media flows through some tubing on its way out, and the embryos get trapped by a fine filter that lets the fluid out but not the embryos.

When the flush was finished Kevin transferred the embryos to a dish so that he could find them through a microscope. He then pulled each embryo out separately, putting each one into an individual straw, for future implant into another cow. I was amazed that you could not only see these small embryos but Kevin can tell just by looking at them if they're good embryos or infertile or degenerative. On the day of the flush two cows were in heat; so they transplanted one of Sleet's embryos into each cow. The remaining five embryos were frozen for later use.

When a dairy farm says it milks a hundred head, it means they actually might be milking around 80 cows, and 20 cows could be dried off. A cow gets dried off about two months before she is due to have her calf. There are more than 100 cows to feed and take care of in order to keep the number of cows being milked at 80 to 100. A dairy cow at Tiffany Farms goes through many stops at different barns before she ends up in the milking parlor. Soon after being born, she is housed in a hutch for the first few months of her life. A hutch is a small wooden structure with hay in it for bedding. The hutches are lined up in a row near an area of grass that is corralled off. In good weather the young calves are let out in the penned grassy area during the day and then put back in their hutches for the night. Sue Tiffany feeds them milk, grain and hay twice a day. Right now there are six calves in the hutches, but six more are due to be born

in the next week; so depending on how many are born female, the hutches will be full again. If a cow has a male calf (bull calf), he is sold at auction, as there is no use for a bull calf on a modern dairy farm.

At about three months of age, the young calves are considered heifers and are moved to a place called the bullpen, which is where the bull used to be kept. It is an area at the end of the calf barn with two inside stalls big enough for five or six heifers to lie down in each stall with sawdust on the floor for bedding. At this stage of their lives, they are weaned from milk and are fed grain, hay and silage. From the bullpen they go to the courtyard and the stalls that are part of the courtyard barn. They are getting to be pretty big by now, and two to three share an inside stall.They are turned out from their stalls to the courtyard in the morning and then put back in them for the night. On average there are six to eight heifers in the courtyard. They are also fed grain, hay and silage. Sue feeds the bullpen and courtyard animals the grain and hay, and Hannah cleans the stalls later and feeds them their silage. From the courtyard

they go out to pasture in the spring and then down to the lower barn, which is across the street and down the road from the house and the rest of the barns. They are then roughly 14 months of age and are ready to be bred by artificial insemination to have their first calves. Dave performs the artificial insemination on 10 to 15 cows a month in order to keep milking the 80 to 100 head.

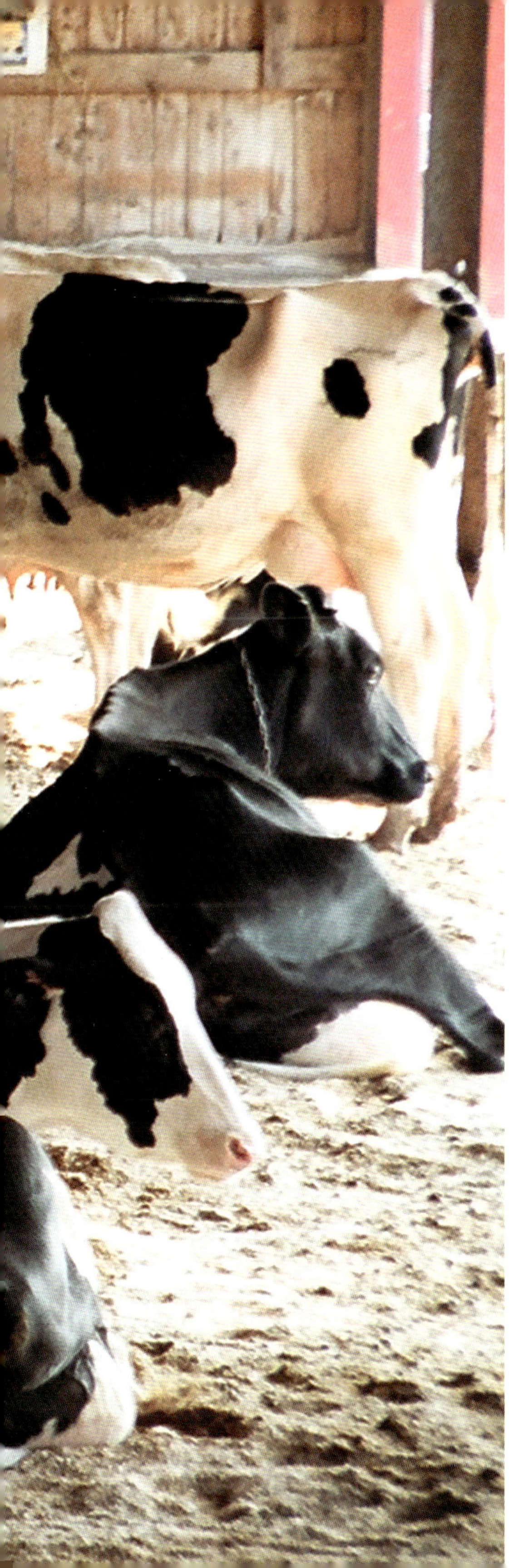

Artificial insemination came into use after World War II and became widely used instead of bulls to impregnate cows around the 1950s. Tiffany Farms began using artificial insemination around the 1960s when Jack's father, Leon, milked Guernseys. Jack told me, "It became very popular here in Connecticut because there were lots and lots of small herds and people with just a few cows, who weren't what you'd call farmers, but they either had to keep a bull or else truck their cows to someone who had a bull and that got to be a pain, so it was well received." A technician used to come to the farm and carry out the procedure, but in those days it was different in that the bull's semen had to be fresh; so the shelf life was not very long. These days, it's frozen and diluted for use at anytime. He said, "It's not unusual for a bull to have 15,000 or 20,000 offspring." He went on to tell me that bulls can be very dangerous, and in the 1950s a local dairy farmer named Willie Peck was killed by his bull and, "my dad was gored by a bull, although he lived to tell about it. Bulls were kind of dangerous. They weren't something you'd play with. Most of them had their horns because they tied 'em up by the horns, but they would cut the ends of 'em to make them blunt." I watched one day as Dave was performing the artificial insemination on a heifer and he explained what he was doing as he started the procedure. "Basically, what I do is thread the rod with the semen in it through the cervix, putting the semen at the base of the uterus." When the procedure was finished he said, "That was it, not quite as fast as a bull, but not much longer." I commented that you have to be pretty calm when you do this so she'll be calm, and he said, "Yep, she was pretty good. Sometimes the heifers [young cows who've not been bred yet] are back and forth, jagging around. She was standing heat last night…we generally breed 12 hours after standing heat." I asked what standing heat was and he said, "Standing heat is when one cow is jumping on another," which means she was being mounted by another cow. They watch for that. I asked if the cows exhibit any unusual behavior when they're in heat and he said, "They generally get very friendly." I asked, "With each other or with people?" He said, "Both." Sue has told me to watch my back when I help her feed in the morning because a cow that was in heat climbed on her back once.

When the vet, Kevin, comes to the farm once a month to check pregnancies and do other vet work that might need to be done and confirms that a heifer is indeed pregnant, she goes back across the street up to the field by the road where all the pregnant cows that are being dried off are. The cows that have been in the milking string and are pregnant go up to the dry cow lot two months before they are due to calve. Jack had a sign made for the dry cow lot that says "Ladies in Waiting" simply for a bit of humor. For years I thought the "Ladies in Waiting" sign on the wall of the feed trough meant that the cows were waiting for their food. It turns out that it means the cows are pregnant and are waiting to have their calves. I found this out when I asked Jack one day where the dry cows are kept. I really didn't know what the term "dry" meant, but I'd heard him use it. He said, "Well, those are the ones up by the road. That's what the 'Ladies in Waiting' sign means, they're pregnant." He went on to say, "It goes back to the Victorian times. That was a fairly common phrase back then. If you were pregnant, you were a 'lady in waiting' rather than pregnant." About two weeks before a "lady in waiting" is due to calve she is moved to the milking herd in the loafing barn so that Dave can keep an eye on her and assist with the birth. There are probably 150 animals at the farm to be taken care of and fed in order to milk 80 to 100 head year 'round.

LADIES IN WAITING

The first person up and about working at the farm is Dave, who does the morning milking starting at 5:00 AM. As he's rounding up the cows from the loafing barn to go into the milk parlor he sanitizes the milking equipment. Once the equipment is sanitized and in place and the cows are lined up, he calls them in by whistling. There are four elevated stalls in the milk parlor, and as one cow is finished milking she's let out and another comes in. A robust whistle is usually all it takes to get a cow to come into the stall, and it seems that all the milkers know how to whistle loud enough, although sometimes they need to call the cow in by name. Cows generally line up to be milked in the same order every time. Some have strong preferences for certain stalls, too. Once the cow is in the stall, Dave takes a pre-dip anti-bacterial liquid and puts it on each teat. He lets the dip set for 10 to 15 seconds and then dries it off with a clean paper towel. Then the machine is hooked up to each teat and the milking takes about five to ten minutes. The milking machine comes off and another dip, called a barrier teat dip, is put on to help protect against any bacteria out in the loafing barn and to prevent mastitis. The entire milking takes about four and a half hours in the morning and slightly less time at night, owing to the fact that there's a shorter period of time between the morning and evening milking than there is from the evening to morning. On the large dairy farms where they milk 1,000 cows, they milk around the clock, I've been told. Something else that is a drawback to milking so many cows is the question of where to put all the manure. A thousand cows can put out a lot of manure, and the farms are required to bury it or cart it off the property as too much manure is harmful to the balance of the soil. There's a lot to be said for keeping the operation to a modest size, as there is not as much waste on a smaller scale farm. Anyway, the main issue with milking as a way of life is the fact that the cows have to be milked on a regular basis, twice a day, seven days a week. At Tiffany Farms Dave does the morning milking and Hannah milks in the afternoon, starting at 4:00 until 6:00, when Dave finishes up for her so she can get dinner ready at home. Jack milks on Sunday morning until 9:00 when John, his son, finishes up. Kenny fills in when he is needed. The schedule is relentless, but Jack was telling me that, "When I was a little kid, Frank Walden told me that he worked for Ed Bill for twelve years without a day off. And when Dad started milking here, the herdsman got two days a month off. That was considered liberal. The herdsman's job was a seven days a week job, and, of course, you had chores and you had the middle of the day off, but you were still responsible for milking and feeding the cows seven days a week unless you were deathly sick. Things were different back then, people were glad to have a job in those days. There wasn't any unemployment or welfare."

When the Tiffanys started milking in 1941 they had 28 cows and milked the cows in about an hour, but the cows didn't give as much milk then as they do now. Jack explained the milking process: "We had two rows of stanchions out here, with the cows headed out, just like at the lower barn, with another row on the other side, and you started on one end and worked right through to the other, carted the milk by pail to the milk room, weighed the milk, strained it, put it in cans, and then put it in the cooler. A truck would come by to pick up the cans of milk and leave some empty ones." About a dozen or 15 farms in the area in the 1940s operated this way and on this scale. Some of the farms were Raymond Farms, Tiffany Farms, Firgelewski's, Harding's, Sankow's and Machuba's. Not all of the farms had their milk picked up at the farm, and the farmers had to take the cans of milk to a stand at the school on Route 156 to be picked up. When the farms were forced to switch to bulk tanks for milk storage prior to the pick-up a lot of the farms went out of business because the tanks were very expensive.

In addition to the cows that were milked by machine in the 1950s, the Tiffanys also had about 10 cows that were milked by hand by Leo Dossie down at the lower barn. Jack said, "Leo Dossie used to sit on the grain box until 5:00 AM on the dot. He wouldn't start milking until the minute. If he got his other chores done ten minutes early, he'd wait. He knew that cows were

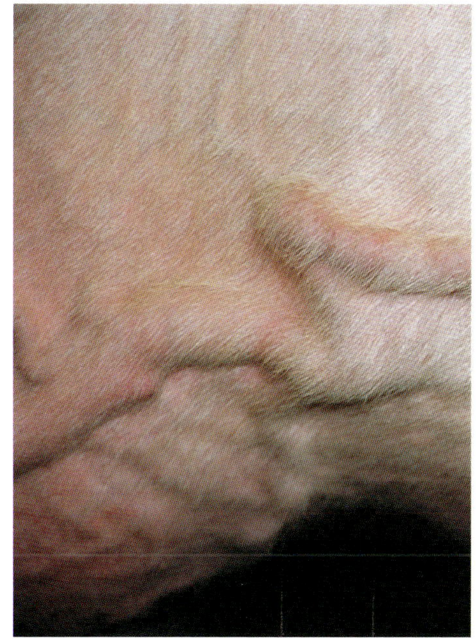

animals of habit and punctuality." I wondered how long it would take to milk a cow by hand and Jack told me a good milker could milk a cow in about five or six minutes. Every morning and afternoon the cows line up very patiently, waiting their turn to be milked. I wondered what happened if a cow that had a full udder didn't get milked, if it would be painful, and Jack told me that the milk would eventually be reabsorbed into the body, but it would be somewhat painful. Farms are run on routines and the routines are all dictated by the cows' needs.

The Hamburg Fair has been an annual event in Lyme, Connecticut, since 1897. This year marked the 103rd fair. Oxen pulls and animals have been major attractions at the fair since its beginning, but the Pony Pull was introduced to the Hamburg Fair by Leon Tiffany, Jack's father, in the mid-1970s. After having a mild heart attack in 1970, Leon turned the farm over to Jack and Lee, Jack's brother, and shortly thereafter developed an interest in ponies. He and some friends, Ted Emerson, Wallace Prann, Merritt Comstock and Henry Troger, went to Maine to the Fryberg Fair where they saw some ponies. Leon and Wallace then went to Pennsylvania, where they bought their first ponies.

Jack told me, "They started with the little Shetland ponies. Goldie and Betsy were his first pair and they were driving ponies, not pulling ponies. And then he got some friends together and they formed the Connecticut Draft Pony Association and he was the first treasurer. They sanctioned pulls around the state and he used to have pulls here at the farm. He had a pair of spotted

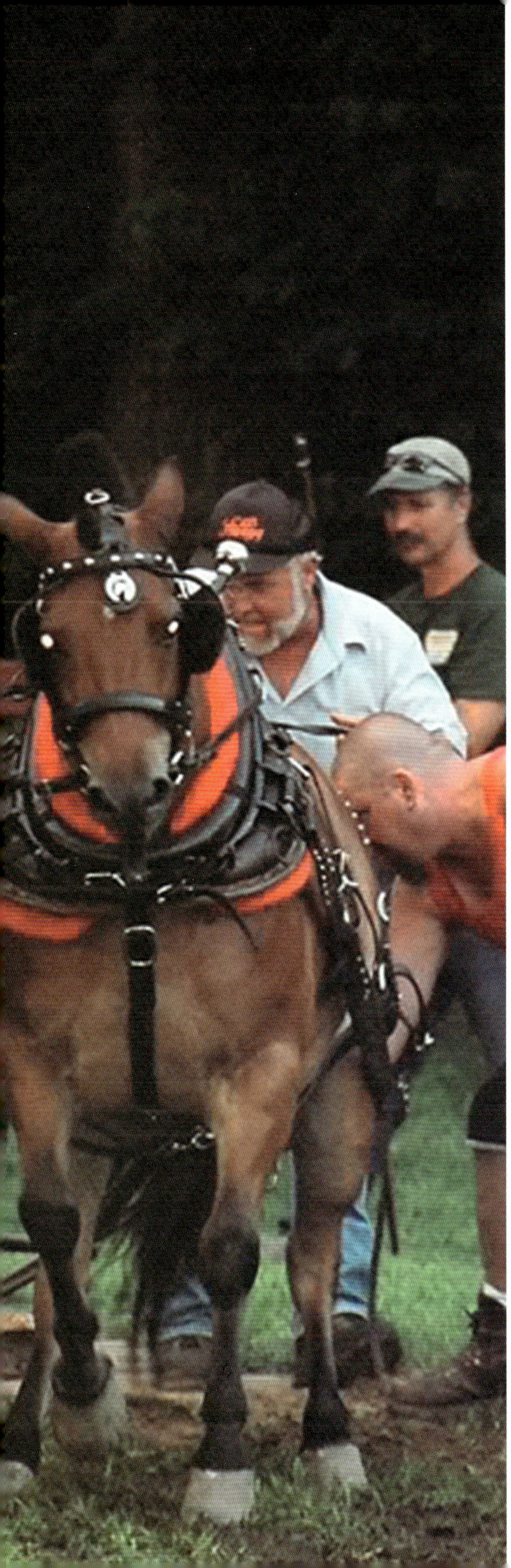

ponies named Sonny and Stripes, and Sonny and Stripes really cleaned up. They were the New England champs. Of course, he did it for the fun of it, not for the money. There was very little money in it in those days, kind of like the ball players. He wouldn't really make expenses hardly when he first started pulling. He bought a new truck, built a new body, two fancy harnesses, he went into it full time. He would drive the truck and meet the teamsters at the fair. Sometimes the teamsters would stay here at the farm and go to the shows with Dad. In the mid-1970s he introduced the pony pull to the Hamburg Fair and he was chairman for many years. He'd always take the 40-20, that's the tractor with no cab, and pull the drag [concrete blocks] back. He'd build a stand for the judges, have it decorated with corn stalks, ribbon and bunting. Nick Naples was the announcer for many years. Then in Dad's last years, they kept him on as honorary chairman; Gary Reynolds does most of the work [now] and I pull the drag back. I'm not sure whether it was the first year after he died or shortly thereafter, I got the idea of having a trophy for sportsmanship, because he admired sportsmanship. It's not necessarily for the team that pulls the most weight. We've been giving the trophy ever since." The trophy is called the Leon H. Tiffany Hamburg Fair Pony Pull, Best Teamster Award, and was first awarded in 1982 .

Leon died in 1982 and some of the ponies outlived him. Jack told me, "Molly Jewett, you know, loved animals, and she took some of the ponies when they were too old to pull and kept one of the team for many years after Dad died. He kept the retired ponies Sonny and Stripes around here pulling the manure spreader and jumper sleds of wood out of the woods. Ponies don't pull forever. They pull for four or five years on average."

RANDEL

Watching cows in the open landscape, slowly eating grass or chewing their cud, can lead one to believe that life on a farm is a more simple way of life than the one most of us are leading. In reality, farming is every bit as high-tech as the rest of the world. The milk tester, Zoanne Toth, who comes to the farm once a month to test the milk told me that she got into the job to get away from a high-pressure management job and wanted to do something more laid-back. Her job as a milk tester may be less stressful than her previous job, but she carries her own computer to the farm to enter data into and she often works seven days a week. Her territory is Rhode Island, parts of Massachusetts and parts of Connecticut.

Milk testing is an important part of farming in that it provides the farm with information, such as how much milk each cow is producing in a 24-hour period and how much butterfat and protein is in the milk. Farms not only get paid for the volume of milk the cows produce but also for the butterfat content, which should be around 3.6 percent, and the amount of protein, which should be around 3.0 percent. Cows can be bred to produce more butterfat or protein, so it is helpful to know which cows are high producers. Then, once the cows are identified as being good producers of butterfat, protein or volume of milk, Dave and Hannah try to improve the herd by flushing these cows and implanting the flushed embryos into the cows that aren't high producers and have them carry the offspring of the high producing cows.

The farm belongs to an association called the Dairy Herd Improvement Association, and the samples of milk taken from each cow when Zoanne is there testing are sent to a lab in Ithaca, New York. One of the tests is for somatic cell, or the health of the udder. Zoanne told me, "If I find a cow that has a very high somatic cell count, that cow is probably sick or coming down with mastitis and needs to be watched or treated. If the farm has a high somatic cell count in their tank they get hit for quality and the price they get paid for their milk goes down." In addition to the tests run for butterfat or protein content and somatic cell count, there is also a test called MUN, or Milk

Urea Nitrogen, which tells the farm if there is too much protein in the feed. The results of the tests are then sent down to DRMS, or the Dairy Records Management System, in Raleigh, North Carolina. Jack told me there are only two or three processing centers in the whole United States. Hannah told me, "DRMS puts all the information from the lab tests done in New York and all the information that we provide on our heifers and cows on computer, and we get back these computer-generated sheets. They send it up to us and I have my computer online at home. I can upload it and get all the information by computer. We get back a couple of pages that have the cow's name, when she calved , how many pounds of milk she had this time, how many pounds of milk she had last time so you can see the difference, her percent butterfat, her percent protein, pounds butterfat, pounds protein, how many pounds she's got this lactation to date and what her estimated milk is. And then there'll be heifer information, with a list of all the heifers and when they were born. And then you'll get these action sheets which will tell you when the cows that have just calved should be bred again. We usually wait 90 days after they calve to breed them back."

When the cow comes into the milking parlor Zoanne puts a meter on the milking equipment that tells her how much milk (in pounds) the cow produced in that milking, and she records that number on the lid of the sample container. Cows are milked twice a day, so there are two testings per 24-hour period, and she records both of these numbers on the lid of every container. Jack told me, "When we first started milking, this was all done on the farm. The tester used to come, of course then there were hundreds of more farms that belonged to the program, but most of them were all small, had less than 50 cows. And the tester would come and he would stay overnight. He would do the night milking and do the morning milking. There was this Babcock test where they put milk into these little vials. The vial had a long, thin neck that was graduated and the milk was centrifuged in a hot water bath, and the fat rose to the top. You took a pair of calipers and measured it and it would tell you the amount of butterfat. They didn't measure protein in those days, just butterfat, but it was amazing how accurate that was. It was all done manually with logarithmic tables once a month. Christ, he'd be here until 11:00 at night doing 30 cows."

Once Zoanne is finished taking all the samples, she enters the milk weights for each cow into her computer, which Hannah can then take and enter into her computer. Zoanne said she likes to get all her farms on their own software programs so they have everything in their own computer base. She said, "The program tells them when to breed the cow, when it's time to dry them off, when it's time to freshen; and then we do studies on the effectivity of their breeding program. It's a little more complicated than just milking a cow." In addition to her work as a milk tester, she also trains other testers, installs software and gets the software to the farms. Her smallest farm milks 38 cows and the largest milks 600. She said she hasn't lost any of her farms, which is remarkable considering the number of dairy farms that have gone under lately.

It's the end of August and Toby Pinn will be headed back to Bates College in Lewiston, Maine, for her last year as a biology major before applying to veterinary school. She has helped at the farm for the past three summers in order to get experience in working with large animals, which she hopes to specialize in once she graduates. She said she can't remember not wanting to be a vet, and part of the reason for this is that her parents used to bring her to Tiffany Farms to watch them milking the cows and to see the calves when she was a young girl. She also told me she loves the working environment at Tiffanys' and that, "More and more, she would like to focus her studies on bovine medicine, because she doesn't want to see the industry that supports this field die out." In addition to working at the farm in the summers, she has gained veterinary experience from working at various equestrian camps and through internships with vets near Lyme and her college in Maine. At one camp she was the herd manager for about 50 horses and went on call with the bovine-equine vet there. I asked her if she was ever intimidated by the size of the large cows and horses she works around, as she is not a large woman by any means, and she said, "No, not really. It's not that they intimidate me, it's that I sometimes have a hard time getting them to notice me. They might overlook me, so I think I have to be a little louder or a bigger presence. Once I get used to it, it's not really a problem. I know that cows aren't aggressive and you can't muscle a horse around, so you have to learn to work around them. I don't think it matters how large you are with a large animal; you just have to know how to work around them. The lifestyle of a large animal vet is more of a factor for women who want to start a family, because you're on the road all the time." She told me that in her veterinary classes the majority of people are female. She hopes to be able to establish herself in a large-animal practice and eventually have a family as well.

Since this project began almost nine months ago now, I have enjoyed getting to know all the people who work at the farm and find it interesting to ask them what drew them to farm life. Seana Bill has worked at the farm for the past two years, and I asked her how she came to Tiffany Farms and what she likes about the work. She told me, "My brother used to work here and then he went to the Naval Academy. I started working here when he left. We have kind of a small farm at my house with horses and stuff and we used to have cows. I've always gotten along with animals pretty well." The thing she likes best about her job at the farm is that, "the people are really great, nice and laid-back. It's a nice working environment." She is in her first year of college at Avery Point, which is in the area, so she plans on continuing to work at the farm part-time on Fridays, Saturdays and Sundays. Her chores include cleaning out the calf barn and all the stalls in it, scraping the courtyard there and at the lower barn, as well as cleaning and spreading sawdust on the floor of the loafing barn. She told me she worked six different jobs this summer for a total of 60 hours a week. The jobs were all quite different, starting with the work at the farm, babysitting, bussing tables at a restaurant, working two gardening jobs and coaching the Middle School Learn to Row Program for crew. That is a full life.

Sue Tiffany wears many hats around the farm. Her day starts at 6:30 in the morning when she heads out to feed the heifers and young calves, repeating the task in the afternoon around 3:30. In between these chores she does a number of things that keep the farm running smoothly. It is said that women on the farms are the glue that holds the operations together, and that is certainly the case here. She babysits Hannah and Dave's two children, Erica and Lauren, so that Hannah can do barn chores and milking. She arranges for the extra hands that might be needed for haying and calls anyone that needs to be called for deliveries of supplies or repairs. One of her biggest responsibilities is paying the bills, which is becoming more of a challenge to make ends meet than it used to be. She told me, " When I started to do the checkbook for the farm, which was back in the early 80s, I took over for Jack's mom and dad. I used to be able to sit down, pay the bills with the milk check and automatically save 500 to 1,000 dollars from every milk check and put it towards a piece of machinery. Now I can't save anything." The milk check comes twice a month, which meant she could save up to $2,000 a month then. She said, "The cost of everything was about a quarter of what it is now and the price we got for the milk was about the same as it is now. The cost of grain has about quadrupled and labor is very high, and taxes, real estate taxes are high." She went on to give me a breakdown of where the money goes, "We get two milk checks a month, on the 5th and on the 20th. The one that comes on the 5th is called an advance. It doesn't have many deductions from it. Out of that check, a fifth of that is for payroll, our regular employees. We try to give at least a third of that check to the grain companies for the grain. About a tenth would go to a loan we have with First Pioneer Farm Credit and about another tenth would go to Social Security and withholding taxes. It has to be paid to the bank, and on the 20th about a tenth of that check would go to utilities and electric for the farm. The rest, if there's any balance, we've got to spread out between all the different vendors from whom we buy supplies. It's hard to stretch it." The milk check went up a little this month, probably because of school starting up again, which increases the demand for milk. When the checks are low, the federal government subsidizes their income.

Most small to mid-size farms are losing money every year. Some facts put out by the Connecticut Farmland Trust, which is interested in preserving Connecticut's farmland, are:

• Across the United States over 360 acres of farms and forests are converted into subdivisions and shopping plazas every hour (6 acres every minute!).

• Connecticut loses 9,000 acres of farmland a year. In less than 20 years Connecticut has lost over 20% of its farmland. Connecticut has lost a greater percentage of farmland than Maryland, New Jersey, and Massachusetts, which have high population densities and the same development pressures that Connecticut faces.

• Connecticut's farming operations contribute $2 billion annually to the state's economy.

• Connecticut has some of the most fertile farmland soils in the world, especially within the Connecticut River Valley.

• An American Farmland Trust study of five Connecticut towns found that for every $1.00 of tax revenue raised, it cost $1.13 to provide public services for residential land, compared to only $0.34 for farmland. (Durham, Farmington, Hebron, Litchfield, and Pomfret).

In addition, the Farmland Trust suggests some things the ordinary citizen can do to help protect Connecticut Farmland:

• The State of Connecticut lags far behind most other Northeast states in its commitment of public funds for farmland preservation. Let your state legislators know that farmland preservation is an important public priority.

• Many towns have taken steps to protect farmland as well as becoming more "farm-friendly." Let your town officials know that farming is vital to the future of your town.

• Support the work of the Connecticut Farmland Trust, which is securing agricultural easements from farmers and other landowners to protect farmland for future generations.

Autumn is thought to be the time of harvest, getting the stores of food in for the winter. In our modern age those of us who aren't farmers don't recognize the harvest as readily because we can go to the grocery store anytime to buy pretty much whatever we want, whenever we want. If a certain food is not in season here, it is somewhere else in the world. On the farm, however, the season's last haying is being wrapped up and the chopping of corn for the silos will be beginning soon. The fields will be put to bed for the winter with the planting of rye grass that will hold the soil together until the spreading of manure begins again in the spring. The frenetic pace of the spring and summer field work will be over, and the constant watching of the weather for haying purposes can take a rest. The colors of goldenrod and fall foliage are ablaze, as if to say, "Before we go, we'll put on our best show to keep you going until spring." ■

Autumn

The Tiffanys have a small herd of beef cattle that they raise for their own use. Two of the beef cows had calves recently. One of the cows that is mostly Shorthorn had a really beautiful baby that is speckled roan from the waist back with solid-colored red legs, chest and head. The mother of this calf has always been my favorite cow to paint at the farm. Erica, the Tiffanys' granddaughter, told me she is her favorite cow too. There's something about her face that makes me laugh, and I titled the last painting of her "Eat More Chicken" after the commercial that was on TV for awhile. The other cow that had a calf recently was an all black cow with horns. She was bred by artificial insemination with Belted Galloway semen. The Belted Galloway breed is a black or dun (brown-colored) animal with a white band around the middle. The nose is black and they are naturally polled, meaning they have no horns. The white belt around the middle can be of various widths but generally runs from the shoulder to the hips. They are often referred to as Oreos, as in the cookie. The Belted Galloway was brought to the Galloway region in the south of Scotland by the Vikings, and the breed is thought to be the oldest breed of cattle in Britain. They are a hardy breed that can live in harsh winter climates, thanks to their double coat of hair, which resembles that of a buffalo. The coat is comprised of a long overcoat, which sheds the snow and rain, and a soft undercoat that traps the warmth. Since the calf born to the all black mother was a mixture containing Belted Galloway, her white stripe around the middle does not go from the shoulder to the hips, but is a tiny, jagged stripe instead. That's what I like about the beef herd at the Tiffanys: they are not purebred cattle, but mixtures of various breeds, such as Shorthorn, Hereford, Charolais and Belted Galloway; so the variations in coloring and stripes are a treat to the eye.

By the end of September, the second cutting of hay is being wrapped up in the last remaining field. From May to October it has been a relentless schedule of transporting machinery such as the mower-conditioner, tedder, hay rake and baler from one field to another, constantly watching the weather in order to make sure there is enough sun between the mowing and baling at each field. Although I imagine the field work is something that can be very pleasant on a beautiful day, it is still a lot of work. It must be somewhat of a relief to put the baling equipment away for the winter. The haying season this year started out very iffy because of all the rain, but ended up being all right in terms of the amount of hay baled. The quality of the hay baled in the spring, or the first cutting, was mixed as a lot of the hay got wet between cutting and baling and will be used for mulch hay. The second cutting of hay is usually of a better quality than the first cutting anyway.

Now that it is the beginning of October the push is on to get the corn chopped and into the silos before we have too much of a killing frost. There is a fine line between letting the corn dry out enough before chopping it and having it dry out too much from a killing frost. It takes about four to six weeks to chop the 100 acres of corn planted in various fields, so the corn that is chopped last is bound to get more frosted than the corn that is chopped first. Corn is chopped much the same way that grass is chopped. The chopper is pulled by a tractor and the corn gets cut close to the ground. It gets fed into the chopper, stems, leaves and ears, and after it's chopped it gets fed into the truck that drives beside the tractor and chopper. Then it's blown up into the silo from the self-unloading truck. Jack told me about how corn was harvested when he was a young boy: "When I was a kid, it was my job to ride the horse and a man walked behind us with the cultivator. He cultivated two rows at a time. Of course you only grew maybe 15 acres of corn, whereas today you grow 100. It was planted by hand, cultivated by hand and harvested by hand. You had hoes or corn knives and it was all hilled then. It wasn't planted in rows. A hill is where five stalks of corn are planted close together. You took a tool and marked thirty-inch rows one way and then you marked thirty-inch rows the other way and where they crossed you dropped five kernels; and then you'd take your shoe and scuff the dirt over and clomp it down. And then you'd go along, when you harvested it, you'd go along and wrap your arm around those five stalks and take a hoe and cut 'em off and then drop it down and then you'd go along with a wagon and pick up those stalks and cart them to the silo and run them through the blower at the silo." He was fifteen years old when he helped harvest corn this way, which would have been 1947. They started harvesting corn in a different way in the

1950s when the corn binder was first used at the farm. "The corn binder would cut the corn off, take a string and wrap it around the bundle just like a hay baler. Then the bundle was either dropped on the ground and picked up, or there was an elevator on the back of the corn binder that dropped them into a wagon behind the corn binder, where a man would stack the bundles. You brought the wagon up to the silo and you had a stationary chopper and chopped it at the silo. Then they came up with the field chopper and that was a really big thing, but now they've got these big, self-propelled choppers that don't need a tractor and they cut five or six rows at a time. And they have these big, huge dump trucks and they blow it into a dump truck and back up and dump it into a trench."

Jack went on to tell me, "Corn silage is the main source of roughage in feed for dairy farms in Connecticut and indeed for most of southern New England. The reason being that you can raise more TDM, that's total digestible nutrients, with corn per acre than with any other crop. In other words, you can feed more cows per acre with corn silage than you can with alfalfa or grass. There's a lot of energy in corn, in the ears. Some farmers even grow enough corn so that they pick just the ears and can have both corn silage and corn as a grain to supplement the silage with. They don't have to buy as much concentrate. Corn has some drawbacks in that it is very susceptible to wind damage. You haven't got to have a hurricane, a heavy rain and strong winds can do a lot of damage in a few minutes to corn and, of course, a hurricane is extremely devastating to corn. Hail will damage corn, although we're not in an area where you get a lot of hail. Through plant genetics they have bred corn to be resistant to a lot of its natural pests. And they've developed new varieties that grow half again as much tonnage per acre as they did when my dad was farming. Corn is a major crop for most farms. There's always a rush to get it in on time, because it's an annual and frost kills it. And when it gets badly frosted, it dies and gets dried out quickly. It's very difficult to store. It will keep in an upright silo much better than it will in a pile. Most farmers who grow a lot of corn store it in piles on the ground or in trenches, and they've got to have it fairly high in moisture." The big silo at Tiffany Farms is 24 feet in diameter and 80 feet tall and holds close to a thousand tons. That silo gets completely filled and there is more corn silage that gets put into a smaller silo also. Jack said that this was a good year for corn because we had so much rain. The heavy amount of rain we had in the spring delayed the planting somewhat and some corn had to be replanted because of so much rain, but all in all it was a pretty good year for corn.

Sue Tiffany loves cats and it's a good thing, because there are currently ten of them in the house and five in the barns. Once there were seventeen cats in the house, and one spring thirty-one kittens were born. She realized she had to get them all fixed or she would soon have hundreds. Fortunately, Paul Urband, a veterinarian in nearby East Haddam, who has since moved to Vermont after a stroke a year and a half ago, gave her a break on the cost of tackling the job. Getting this number of cats spayed and neutered is not like taking your loyal dog to the vet, but one by one Sue took them to Paul. He told her to come with them whenever she could catch them, even if it was midnight. She finally got down to the last cat, but this cat was so wild and unapproachable that it would have to be caught in a cage with bait. This method doesn't always work with the really feral cats because they are so skittish. Anyway, Jack rigged up a cage with food in it and the cat did go in, and he pulled the door shut from a distance with a string attached to the door. They called Paul but found he was out of town. In the end, Kathy Clarke, a veterinarian in Old Lyme who does a lot of work with stray and feral cats, took her and spayed her. The job was finally done. Seeing to it that every single cat, both inside and out, was fixed was quite an accomplishment.

Corn silage is just one of many components that make up the feed for the cows. Jack Tiffany puts together the ingredients for the morning feeding, which takes about an hour in all, and Hannah, his daughter, puts together the night feeding. I asked him to explain what goes into the mix and he told me, " Well, the morning feeding isn't any different from the night feeding. Pretty much all dairy farmers feed what they call a Total Mixed Ration, TMR. You mix together your roughages, which would be hay and silage, and your concentrates, your grain and commodities. In the old stanchion barns, you generally fed them separately, but scientists have decided that it's actually better to mix everything together. It's easier for the cows to digest. Your basic roughages are grass or alfalfa silage and corn silage and dried long-stemmed hay, which adds digestible fiber, which makes the cows chew their cud more. And when they chew their cud, you get better utilization of the feed and it helps raise the butterfat test a little bit. You have to determine a feeding program and decide how much milk you want to get out of the bunk. We're feeding for 70-75 pounds of milk out of the bunk, and when you determine how much milk you're feeding for, you run it through a formula and that tells you how much net energy you've got to have in the feed ration. So, you kind of work backwards. You know what you want to produce and then you determine how much corn silage and how much haylage you need. You send your forages away and have them analyzed. That's done periodically. That tells you what's in the feed. You do that for the corn silage and the grass silage and the hay and then you determine how much of the grain or commodities you want to add. Now, there are all kinds of commodities you can purchase and they all have a pretty common analysis, not between themselves, but between various loads. It isn't like corn silage or hay, where you have good silage and poor silage. Corn meal is generally corn meal. It'll test pretty much the same no matter where or when you buy it."

He told me there are dozens of commodities that you can feed and some of them are regionalized, which cuts down on the cost of transporting them to the farm. The farm uses different commodities, depending on the price and what they're trying to achieve in the bunk or in the mix. He said, "We generally use corn meal year-round, and soy meal, which is very high in protein and brings the protein content up. And then for digestible fiber, we will feed beet pulp or cottonseed, and the alfalfa we bring in from the far west. We have a stationary mixer, which is on scales. It's just like baking a cake, instead of measuring by cups or spoonfuls, it's measured by pounds. After it's mixed together we put it in a truck and take it out to the bunk. Some of the big farms have what they call a high producing group, a middle producing group and a low producing group, or maybe just two groups. And those groups would be fed different mixes. We feed all our cows the same mix except for the dry cows, and the beef cows get a separate mix." They do supplement a cow's feed ration with more grain if her milk output is greater than 70-75 pounds. This is done by using a computerized system at various feed stations that dispenses grain to a cow with a transponder around her neck that is read like a bar code at the supermarket. The cow puts her head in and the transponder on the chain around her neck is read and the amount of grain she is supposed to get periodically throughout the day is automatically dispensed. The data that the farm receives from the results of the milk testing that is done once a month tells the farm which cows are producing milk in excess of the average, and these results are programmed into the computerized feeding system at the farm. Accurate information is very useful for the proper feeding and utilization of the feed, both nutritionally and economically speaking.

MILK AND MILK
3067
Connecticut
V·60384
Trailer

The culmination of all the careful breeding, relentless field work, barn chores, maintenance of equipment and buildings, payment of bills and proper feeding of the cows is the end product, the milk. The milk gets picked up every other day by tanker truck and is taken to Guida Dairy in New Britain, Connecticut, where it is processed into many different products. Tim Heim, Sr., who picked the milk up, explained the process to me, "You measure the milk, you smell the milk, you visually look at the milk and you take a sample of the milk. You have to let it agitate for five minutes. You measure the milk and then you correspond the stick reading with the chart and that tells you how much milk is in the tank, and then you just pump it into the truck. The average pick-up here is about 11,000 pounds every other day, and I pick up at three other farms, which will give me a whole trailer load of about 54,000 pounds. The tank holds 65,000 pounds and is divided into two compartments. The reason for the two compartments is that if you were to climb a mountain or something, all the milk would rush to the back and you probably wouldn't make it up, so they split the tank into two compartments to keep the weight over the wheels. The trailer truck is not refrigerated, but is very well insulated. We pick up the milk at 38 degrees and it will stay cold for two days."

Tim told me he used to have his own trucking company and that he's done this all his life. He got his license to do this when he was 13 years old. He said, "I couldn't drive when I was 13, but my mom would drive and I would pick up milk before I went to school, pick up a couple of farms, get on the work program, get out of school early, come home and pick up at a couple of more farms. Then we'd deliver the milk to the dairy." He eventually went to work driving for a company instead of having his own business, because the farms where he was picking up in the Ellington, Enfield area were disappearing, so it wasn't worth his while to stay in business for himself.

Milk has to be picked up in all kinds of weather, no matter what, and he told me he has picked up milk in every blizzard there's been since he started. He told me he picked up milk in a tornado over near Bradley Field one year. He said, "I was over there picking up milk and I had just left a farm when it hit. The next day I went back and I found out that it ripped the roof off a barn and killed a couple of cows." Since milk is a perishable item, it must be picked up, even on holidays. He told me he went eight years picking up milk without a day off, starting when he was 16 years old until he was 24. He said his father decided he was entitled to a day off.

Once the milk is picked up from the farm, it goes to Guida Dairy in New Britain, Connecticut, where it is processed into many different products. Al Guida III, the current president, told me how his father, Alexander Guida, Jr., bought the Seibert Dairy in 1947. He told me, "My father started in business in Middletown. He was one of 13 children and lived on a dairy farm. He came to New Britain and started his own home delivery of milk just after the Depression and built up his business. He was buying his milk from Highland Dairy up in Hartford, which was then purchased by Hood. Hood called in the sub-dealers and said milk was on a ration and they didn't want sub-dealers anymore, and he basically had two months to either find an alternate source of milk or sell his business. Being young and full of ambition, he and his brother, Frank, went to Mr. Seibert, who was an elderly person without young people in the family, and Mr. Seibert said 'I don't want to put up your milk for you but you can buy the company' and that's how it all happened."

I can't get over how much the business of supplying milk has changed from then to now. Guida Dairy is an extremely complex system of computer-operated machinery that does everything from testing the milk to making their own plastic containers, packaging, processing and moving the milk from plant to truck to store. About five or six years ago, Dean Foods was in the process of acquiring all the regional, family businesses, saying name your own price, name your own terms. Al told me, "We said, 'No, we believe there's a place for a viable independent in the market place. There should be a choice, but we still have to realize that we have to meet the large business' commodity pricing,' and we always have and whether we made the right decision or not, I guess history and time will tell. We've had a tremendous growth period in terms of consolidation. The roll-ups of other dairies and the closing of other dairies in New England and people being treated differently than they had been and our local presence has contributed to our growth. We went from $70million to about $130million in basically two years. We had to double our facility and

sink an awful lot of money into it along with the growth of well over 300 employees, and we changed from a level-two company to a level-five company almost overnight and have had to learn to do many things differently, and we're still in that process. It's a very difficult, painful process."

I was given a tour of the facility by Hugh O'Hare, the product quality manager and assistant plant manager at Guida, and he explained what happens to the milk once it gets to them. "When the milk is loaded onto a tanker," he said, "each farm is individually denoted as to how much they contribute to the vehicle based on the stick reading or the measurement on the farm tank. That milk is transferred to the tanker and it's brought to our facility once the truck completes its pick-ups at various farms. Before it's unloaded here, it's screened for bacteria counts and antibiotics, which is a concern, as you know. Any animal can be treated for a disease or inflammation by antibiotics. Those animals are supposed to be restricted from milking, in other words, they milk that product onto the ground or destroy it and they can't put that into the commingled supply, because even five parts per billion of an antibiotic residue is enough to cause a human reaction in some people with antibiotic sensitivities. So, what we do here when the product arrives is we're mandated by law to screen all those loads for antibiotics. If the milk delivered doesn't meet our criteria as far as bacteria or antibiotic counts, then that truck is rejected back to the cooperative and the co-op has to get rid of it. Each individual farm is sampled as well as the whole truck, so if we find a problem with a truck we can pull the individual samples for the farms and find out which farm is the source of the problem. Once the milk is unloaded, we can use it for several different processes. One would be, we'd take some of the milk and put it into a silo and separate it. When we separate the product, we turn it into skim and cream by centrifugal action. When it's separated, we take a portion of the product, take it out of the silo, and we run it through what's called a separator. A separator is a big, stainless steel series of disks spinning at 5,000 RPMs and what it does is, by centrifugal action, it splits the cream and skim fractions of the product to water-based and oil-based particles. They're different weights, the fat part of milk is lighter and tends to float to the surface. You know, the old days before homogenization, you saw the product with cream on the top, that's because the fat portion is lighter than the water based portion, so that would separate itself. So, now we take that skim and cream and we put it in separate tanks and we hold that for later use. Now a portion of that skim is going to go back into that original silo or to another silo to take that milk that's between 3.6 and 4% fat and bring it down to 3.25% fat, which is the standard for whole milk. We don't want our whole milk one day tasting really rich and then another day, it's weak. It has to be standardized, so that it's consistent , day after day. Once the product goes through the process of being separated and standardized, we'll test that product again and then we'll bring that up to a blend area to make it into other products, flavored milk products, or bring it directly to our pasteurization units, of which we have three. Those units will take that product and process, pasteurize, heat treat it, and homogenize it on the fly at about a rate of somewhere around 5,000 gallons an hour. Then the product will go to a storage tank to wait further distribution to packaging, and also to be tested by the lab before it goes on to the next step. Once it goes to the next step, packaging, whatever packaging machine that it's assigned to , it's tested again for post-pasteurization contamination or bacteria that causes shelf-life problems or spoilage. We have nine different packaging lines in this facility, and some of those lines have multiple sizes within the line. For example, a filler could do a half-pint carton or a four-ounce carton. We make 180 different items. We're processing up to 180,000 gallons a day. Some weeks, it's three to five million packages, finished goods. So, it's a high-volume operation."

Acknowledgments

There were many people who contributed in various ways to the completion of this book. John Friday, my husband, edited every single word three or four times and helped with my hopelessly haphazard punctuation. He encouraged me when I had doubts, and I could not have done the book without his help and support. When I first thought of doing the project I got in contact with John Hostnik of Greenwich Publishing, and he directed me to Clare Cunningham to design the book and Scott Bolliger of Wallace Andrews to print the book. His recommendations in both cases were a perfect match. Clare Cunningham was a true godsend, as she knew exactly what I was after, and her artistic eye and mine are almost one and the same. I looked forward to meeting with her and arranging the photographs and text, which was always done quickly and efficiently with some laughs thrown in. If I had a doubt about which photograph was right for a certain subject, she could spot the one that was usually the best out of a group. Scott Bolliger recognized how important the book was to me and was also very encouraging in the beginning stages and kept me focused on what to do towards the end of the project. A friend of mine, Veva Crozer, was also a big help in the beginning stages when I knew nothing about doing a book and was truly flying blind. She put many of my photographs on her computer and showed me what they would look like at the size needed for a book of this type. She also designed the invitation for the book signing at the farm and is always enthusiastic about any project I happen to be working on that she can help with.

The camera used to do the photography is a Canon EOS Elan 7, which was a Christmas present from my husband, John Friday, and was bought from Village Photo in Old Saybrook, CT. I knew very little about taking pictures and was not sure if I was up to the job that I wanted to do, but Joe Lambert, the owner of Village Photo, assured me that this camera was up to the task and he was right. It is an amazing camera with more bells and whistles than I know what to do with, but it can also be a surprisingly simple camera to use. Joe, along with Dan Buttrey, Skip Hubbard, Christy Cohen, Kara Suhie and Jordan Harris, guided me through the learning process. Dan was particularly helpful in the beginning. He scanned many of the Tiffanys' old family photographs onto discs and restored many of them. When he moved on to another job elsewhere, Jordan took over for Dan and has done a wonderful job of printing images from the book for the show of enlarged photographs to be held at the Cooley Gallery in Old Lyme, Connecticut. About halfway through the project, my camera needed to be fixed, and Joe arranged for a loaner camera for a month, which was just another example of the help they provided, along with the fast development of the film and CDs. The photographs in the book were taken with Kodak Portra 400 UC film.

I was not an experienced photographer or writer before I began the book, and the encouragement I received from the various people I've mentioned so far was enormously helpful, but it was really my interest in and love of the farm that produced this book. At times it seemed as if the book did itself and I was merely trying to catch up with the images and thoughts put before me. Jack Tiffany was an enormous supply of information for the writing and at times I told him he was really writing the book. Fortunately, he likes to talk, so it was not difficult getting the text needed to go along with the images. Sue Tiffany always made me feel welcome and has the same gift for punctuation that John, my husband, does. Dave Shillawski was very helpful in telling me whenever the vet was coming to do something that I might find interesting, and he and Hannah were a wealth of information regarding everything to do with the breeding and milking of cows. John Tiffany's skills in carpentry kept me with something to paint around the farm for the summer months when I needed to be around in order to photograph what might come up in the course of the day. Kenny Stewart was a joy to be around day in and day out, and his sense of humor is greatly appreciated.

I can never thank the Tiffanys and everyone who works at the farm enough for sharing their lives with me for such a long period of time. To have your life examined with a camera and endless questions for a year would have to be tiresome at times, but they never once made me feel like a bother or an intruder. It was a year I will treasure for the rest of my life and, as I said in the introduction, there were many glimpses of humor and joy along the way.

Judy Friday

November 2003

List of Photographs

Winter

Spring

Summer

Autumn